PEOPLE

and the

LAND

To my father David Bell,
and to the memory of my mother
Martha Bell

This publication has received support from the Cultural Traditions Group
of the Community Relations Council which aims to encourage acceptance
and understanding of cultural diversity.

The Friar's Bush Press
24 College Park Avenue
Belfast BT7 1LR
Published 1992
ISBN 0 946872 55 4

Front cover: Castlewellan fair, Co. Down (Green Coll. WAG 1175)
Back cover: Women selling vegetables (Lawrence Coll. R2703)

Farming life in nineteenth century Ireland

JONATHAN BELL
of the Ulster Folk and Transport Museum

FRIAR'S BUSH PRESS

A LATE NINETEENTH CENTURY ASSEMBLAGE OF PEOPLE, ANIMALS AND FARMING OBJECTS OUTSIDE A COTTAGE. (Lawrence Coll. R1786)

INTRODUCTION

The word 'modernisation' has been used to describe the huge changes which have swept Europe, and later the rest of the world, during the last three hundred years. These changes have included the growth of mass production, international markets, urbanisation and rapid developments in technology. All of these affected the lives of Irish country people during the nineteenth century. This book will attempt to illustrate some of the ways in which patterns of farming life were changed in different parts of Ireland, and at different levels of rural society.

The experience of people in the Third World today, shows that 'modernisation' can have unforeseen, and sometimes devastating consequences. This was certainly true in rural Ireland during the nineteenth century. Throughout the century, commercial farmers responded to growing national and international markets. Overall, crop production expanded greatly during the first half of the century and the second half saw a great expansion in pastoral farming and a rise in rural prosperity. This movement towards a fully-developed market economy, however, was not simply a story of even progress. One of the century's greatest social upheavals and its most horrific social catastrophe were directly linked to farming life. In 1800 the Irish countryside was divided into huge estates which could be tens of thousands of acres in area. The landlords who owned these estates were often at the top of a pyramid of subdivision, with strong tenant farmers below them, and most notoriously, 'middlemen' who let out land to smaller farmers, including a vast mass of desperately poor people. There were many variations within this overall pattern. Some landlords attempted to rationalise landholding on their estates, and one even tried to establish a worker's co-operative, but for many this did not make economic sense. The main source of income for most landlords was the rent collected by their agents, and they did not have much incentive to invest money in estate development. Small farmers survived inside the system by various strategies. In western Ireland, for example, groups of families would combine to lease land and divide it among themselves in a system known as rundale. Decisions on amounts of grazing, the movement of livestock, and the allocation of strips of arable land to farmers in the group were all made communally.

Despite this sort of organisation, small farmers and landless people detested the landholding system. Even in Ulster, where a system of 'tenant right' gave farmers some security, eviction, or the threat of eviction, was a powerful weapon used by landlords to ensure payment of rent. Historians have argued about the reality of the threat, but contemporary accounts and folklore archives show the almost obsessive fear it induced amongst small farmers. Besides this landlord-tenant conflict, tensions between large and small farmers and between farmers and labourers were also features of nineteenth century Irish rural life.

During the century a great mass movement developed, aimed at changing the system of land ownership. Various groups were organised which agitated for change. The Land League, founded in 1879, was the best known of these. By the end of the century, parliament had begun to pass a series of land acts which transformed erstwhile tenants into owners of their own farms. Between 1885 and 1903, 60,000 tenant farmers bought their land, and after the Wyndham Act of 1903, another 250,000 gained ownership of their holdings. By 1908 almost half the farms in the country were owner-occupied.

The major catastrophe to strike Ireland during the nineteenth century was the Great Famine, which was triggered off by the destruction of the potato crop by blight (*Phytophthora infestans*) in 1845. Historians are divided on the long-term significance of the Famine, but the horrific suffering arising from it is not open to doubt. More than one million people died of starvation or disease, and the rate of emigration was greatly accelerated. A long-term decline in the population of Ireland began in the middle of the century. In 1845, the population was around 8,500,000. By 1900, it had fallen to less than 4,500,000.

In farming, the decimation of the population meant the disappearance of the tiniest farms. Between the Famine and 1911 the number of holdings of between one and five acres had fallen from 182,000 to 62,000. By the end of the century farms of 15-30 acres were becoming typical. There were many contemporary accounts of these changes, and many solutions proposed to the problems arising from them. From the point of view of this book, the most important writings were those of agricultural improvers, who saw the solution to most of Ireland's problems as lying, not only in land reform and developing the system of trade, but also in changing farming methods. As we shall see, agriculturalists usually had a very poor opinion of common Irish farming practices, and often quite wrongly condemned them. Without their accounts, however, we would know very little of everyday life on the land in the last century.

The earliest systematic attempt to study and change Irish farming began with the foundation of the Dublin Society in 1731. The society had the aim of improving farming, manufactures and 'useful arts and sciences.' Subsidiary local and national societies were organised during the later eighteenth and early nineteenth centuries. The Farming Society of Ireland was founded in 1800, and an Agricultural Improvement Society, founded in 1841, became the Royal Agricultural Society in 1860. The earliest societies organised at county level were in Antrim, Kildare and Louth, and these were followed by Mayo, Roscommon, Fermanagh and Clare.

The effectiveness of societies in improving local farming practices was widely debated, but they certainly did assist in diffusing information about new implements and methods. Most societies organised ploughing matches, and other shows. The journals and papers they produced are some of our most important sources of information on farming during the century.

Several societies attempted to open agricultural schools in Ireland during the century. One school at Bannow in Wexford was opened in 1821 under the direction of the famous Irish agriculturalist Martin Doyle. This school functioned for only seven years, however, and many other attempts were also shortlived. The most famous schools were established at Glasnevin, Co. Dublin, and Templemoyle in Co. Derry. The Templemoyle experiment, begun in 1823, was visited by William Makepeace Thackeray, who praised the combination of book learning and practical work in the fields. Other visitors were not so impressed, however, and many small farmers regarded agricultural schools as just another attempt to produce an efficient tenantry who could afford to pay higher rents.

Other bodies, some grant-aided by government, which aimed to change farming practices, became more important, later in the nineteenth century. Attempts to organise agricultural co-operatives in Ireland, for example, began in the mid-nineteenth century, largely inspired by English examples. Some English retail co-operatives attempted to organise the Irish dairy industry, but this was seen by Irish farmers as an attempt to obtain dairy produce at cheap rates for workers in industrial centres in England. In 1894, the Irish Agricultural Organization Society (I.A.O.S.) was founded to represent the interests of Irish producers. Early leaders of the movement were often very idealistic, dreaming of establishing a 'rural civilization' and a 'co-operative commonwealth.' These grand dreams did not materialise, but by the end of the century the I.A.O.S. was effectively organising the Irish dairy industry and also establishing supply societies which provided farmers with essentials such as seed and fertilisers.

At the beginning of the nineteenth century many writers presented rural Ireland as exotic and archaic. By the end of the century, however, stereotypes arising from this view had become subsidiary to a mass of written evidence which clearly showed that most of Ireland had been profoundly changed, for good and bad, by the forces of modernisation.

THE THIRD EARL OF RODEN AND HIS FAMILY DISTRIBUTING BLANKETS TO THE WIVES OF HIS LABOURERS AT TOLLYMORE, CO. DOWN, 1860s

Many landlords saw themselves as benevolent, engaging in charitable acts, such as the one shown here. The most extreme tensions probably did not exist between large landowners and the poorest tenants and labourers, but between the latter and stronger tenant farmers and small landowners known as 'squireens'. who were often stereotyped as ruthless exploiters of the poor.

BRIDGET O'DONNELL AND HER CHILDREN,
VICTIMS OF THE FAMINE

A FAMINE MEAL

Potato blight first appeared in Ireland in 1845, but it caused even more devastation in 1846, when the winter also was particularly harsh, and large-scale suffering continued in 1847 and 1848. Starvation and disease were worst in parts of the south and west, but few parts of the country escaped entirely. Apart from the deaths, the depopulation of the countryside was accelerated by emigration. Between 1845 and 1855, two million people left Ireland.

ATTACK ON A PROCESS SERVER DURING THE LAND LEAGUE AGITATION

The struggle for land reform sometimes led to violence. Armed police often attended evictions, and rural people attacked those involved in the system. However, the Land League's main weapon was a rent strike.

When withholding of rent led to evictions, the league arranged demonstrations, and also assisted with the defence of tenants in court, and in providing shelter for the evicted families.

EVICTED: A SCENE IN COUNTY GALWAY, 1847
Evictions rose to a peak in the late 1840s and early 1850s. Between
1847 and 1850 it has been estimated that there were 50,000 evictions.

This dramatic illustration shows the emotional power which the issue
of evictions could give to agitation for land reform.

A MEDAL OF THE FARMING SOCIETY OF IRELAND, *c.* 1815

We owe a lot of our knowledge about changes in nineteenth century farming methods to farming societies, which encouraged and recorded improvements. The Farming Society of Ireland functioned only until 1828, but this medal illustrates its enthusiasm in celebrating approved techniques. The medal shows a Scottish swing plough operated by one man, and pulled by two oxen, the apex of ploughing technology for many early nineteenth century agriculturalists.

WORK IN A POTATO FIELD AT TEMPLEMOYLE AGRICULTURAL SCHOOL, 1840s

During the nineteenth century, the notion that agriculture should be taught as a science became widespread. Early experiments in scientific education, such as the Templemoyle school in Co. Derry, were relatively small-scale, but later in the century much of their approach had been incorporated into the national school system, and also into state agricultural colleges and agricultural faculties of Irish universities.

FARMING LANDSCAPE, CO. DOWN
By the end of the nineteenth century, when this photograph was taken, the landscape had become the familiar one of small fields and open hillside and bog. The traces of old fields can be seen in the hilly area to the left of the photograph, however. These were probably abandoned after 1850, when tillage began a long-term decline. (Lawrence Coll. C6152)

THE FARMING LANDSCAPE

The appearance of the landscape changed greatly during the nineteenth century. Fields have been made in Ireland for thousands of years, but the 'patch-work quilt', enclosed landscape which covers most of the country today was largely laid down during the eighteenth and nineteenth centuries. Until the present century, cultivation in parts of the west of Ireland was in unfenced strips which were part of the rundale system. 'Improving' landlords in the nineteenth century often boasted that the removal of the 'ruinous' system of rundale was one of their major achievements.

Nineteenth century field systems are fairly easy to distinguish from older patterns, as the latter tend to be more irregular. The nineteenth century enclosures were often carried out as part of the reorganisation of entire estates and here boundaries are usually clear straight lines of hedges and stones. Species used in hedges varied to some extent. Gorse (whins or furze) and blackthorn were both used, but hawthorn was the most common hedging plant. Nineteenth century farming texts give detailed instructions on planting and managing hedges, and it is still possible to see clear examples throughout Ireland, where the guidelines have been followed. One 'improved' type of field boundary was known as the 'Louth' fence. An earth bank was faced on one side with stones, and young hawthorns were planted at regular intervals between the stones. The earthen bank was grassed over and provided grazing, while planting hawthorn between the stones was claimed to ensure that bushes did not grow too widely apart at the bottom.

Some landlords included an obligation to plant hedges in the terms of leases. The type of hedge might be clearly specified, and sometimes trees (often ash) were included. These were planted at regular intervals along the boundary. Nineteenth century Ireland was very bare of trees. Significant stands were mostly confined to the demesnes of big houses. Few landlords invested in large scale commercial tree-planting, and poorer people were so desperate for wood as fuel, or for constructing furniture and buildings, that mature trees standing unprotected in the countryside might be secretly felled and removed.

Stone walls were commonly built around landlords' demesnes. They tended to be mortared and the stone used was often faced to make the surface more regular. Ordinary farmers more frequently used a dry stone construction for field boundaries. In boulder strewn areas of hilly land, building stone field boundaries was a way to use local material, and clear land for cultivation or grazing. Stone walls vary throughout Ireland with the type of stone used, and the method of construction. The limestone of the Aran islands or the Burren, was sometimes used to produce walls with an open herring-bone pattern, while the huge granite boulders of north-west Donegal or the Mourne mountains were made into much more massive walls.

Less obvious than field boundaries, but as vital, were field drains. Drainage was essential in preparing marginal land for cultivation. At the beginning of the nineteenth century, most small Irish farmers depended on the ancient ridge and furrow system of tilling land to carry off surface water, to an open trench dug near the field boundary. Some large farmers, however, were experimenting with underground, covered drains laid out systematically according to 'scientific' plans. Clay drain pipes were developed in England in the 1840s, and by 1870 were in common use in many parts of Ireland.

Improvers planned not only field systems, but also developed ideas on the lay-out of farm yards. Nineteenth-century Irish farming texts contain many plans for yards, and houses. Some landlords and wealthy farmers did have planned yards built, key elements being stables, byres, storage houses, midden heaps and stack yards or 'haggards'. Improvers were generally critical of the apparent lack of planning on smaller Irish farmyards, but surviving evidence shows that farmers were well aware of the importance of aspect, slope, and the relative position of buildings, middens, and hay or grain stacks. As with so many aspects of rural life, people could give clear reasons for practices dismissed as irrational by unperceptive observers.

FARMHOUSE AND OUTBUILDINGS, SENTRY HILL, CO. ANTRIM
Well-off farmers had large, sometimes elegant houses, with a yard built
at the rear. Many outbuildings constructed during the nineteenth
century are still used today. (Dundee Coll.)

CONSTRUCTING A HAY SHED AT SENTRY HILL
By the end of the nineteenth century, many building materials were in use which are still familiar today. Government grants meant that even small farmers could erect purpose-built outhouses. Hay sheds such as the one shown here were becoming very common in many parts of Ireland by 1900. (Dundee Coll.)

INTERIOR OF A FARMHOUSE, BALLYNAHINCH, CO. DOWN
This photograph shows the kitchen of a better-off farmhouse. The ceiling is tongue and groove, the floor paved, and the walls carefully whitewashed. Other signs of comfort include the clock, the press and the good selection of footwear on the shelf above the door.
(Welch Coll.)

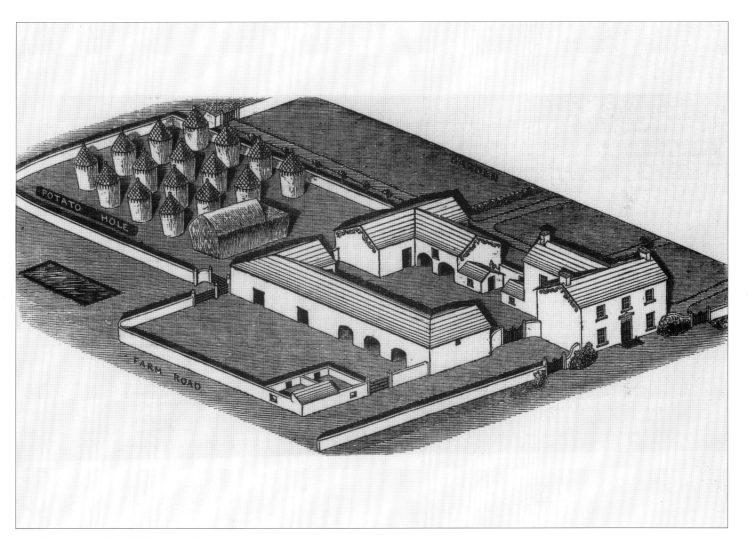

PLAN FOR A FARMHOUSE AND YARD, 1846
This farm was built on the Earl of Longford's estate, Co. Westmeath.
By the mid-nineteenth century, a square farm yard at the rear of the
dwelling house was the plan most improvers preferred. Hay-stacks
were built near outhouses, to make the task of feeding livestock easier.
Grain stacks were kept in the haggard, and threshed either using flails
or by threshing machine.

13

PLAN FOR SETTLEMENT ON ACHILL ISLAND, 1836
During the eighteenth and nineteenth centuries, some landlords
'planted' outsiders on their estates. These people, including German
Palatines, or other protestants, were supposed to act as an inspiring
example to the local tenantry. It is doubtful if any of these plantations
had long-term beneficial effects, and most seem to have aroused only
resentment. This illustration dramatically shows the ideals of farm lay-
out prevalent in the early nineteenth century. A rigid grid pattern was
planned, which ignored the nature of the local terrain.

FARMING IN CO. ROSCOMMON, 1830s
This drawing shows crops of different types growing in unenclosed
strips. Farmers in a rundale settlement were allocated a number of
strips for cultivation each year. This was intended to ensure that each
member of the group had access to land of different quality within the
cultivated area. Improvers claimed, however, that redistribution of
strips led to constant squabbling.

RUNDALE SETTLEMENT, CO. DONEGAL, 1880s

A cluster of houses, (*baile*), whose inhabitants farmed using the run-dale system. Lord George Hill, the landlord of the Gweedore estate in Donegal during the mid-nineteenth century, tried to abolish rundale, and to make his tenants build individual farmhouses in the middle of each newly-consolidated farm. People in Gweedore resisted this, saying that people in the scattered farmhouses would feel isolated. (Glass Coll.)

A BOTHÓG IN CO. DONEGAL, *c.* 1880

A *bothóg* was a dwelling partly built into the side of a hill. There was usually no window and the roof was made of strips of sod or *scraith*. Very poor people lived in *bothógaí* all year, but they were more often used as summer dwellings when cattle were moved to the hills for grazing during summer. These seasonal movements of livestock were often organised communally as part of the rundale system. Unenclosed 'common' land in the hills provided necessary grazing, and moving cattle there meant that the unenclosed arable strips controlled by the rundale community could be cultivated. Stubble and fallow land (*talamh bán*) around the rundale settlement provided grazing in winter. (Glass Coll.)

A DONEGAL INTERIOR, 1880s

In some recent historical writings, it has been emphasised that prosperity improved for many in the farming community. Throughout the nineteenth century, however, travellers from abroad frequently expressed horror at the destitution they observed in parts of the country, and we can be certain that the people who died of starvation during the Famine did not have concealed reserves of food or money. This photograph provides clear evidence of the appalling conditions which many people endured, even in the late nineteenth century.

LANDSCAPE AT CROLLY, COUNTY DONEGAL
This late nineteenth century photograph shows some of the problems faced by small farmers in many parts of the west of Ireland. Before cultivation, stones had to be cleared from the ground, which then had to be drained. Upgrading the acid, boggy soils could take years of backbreaking work. (Lawrence R1380)

DISUSED CULTIVATION RIDGES, CLARE ISLAND, CO. MAYO
The retreat of cultivation from marginal land in the later nineteenth century left traces in the landscape. Old cultivation ridges can be seen on most Irish hillsides, especially at sunset, or when there has been a light fall of snow. They give a good impression of how different the intensively cultivated mid-nineteenth century landscape must have looked from the pastoral landscape of today.

CROPS

The main crops grown in Ireland during the nineteenth century were potatoes, oats, barley and hay. Other crops, such as wheat, turnips and flax were regionally important. Throughout the century there were many general surveys of Irish cultivation methods, and almost without exception these were condemned as archaic and inefficient. The most frequent criticisms were that crops were weed infested, the land was overcropped, and the implements in common use were crudely made and badly used. Agricultural improvers were constantly frustrated by the apparent reluctance of most Irish farmers to listen to their advice. Some new implements and methods were quickly adopted, but other long standing techniques, were kept in use.

What Irish farmers understood, and what many improvers did not, was that 'unimproved' or 'common' cultivation methods were as closely interlinked as 'improved' methods. If one part of the system was changed, other fundamental changes also became necessary. How land was drained influenced how the ground could be broken up for tillage. For example, unless land had been systematically drained with underground pipes, ploughing the ground flat aggravated drainage problems. If, however, land was drained by the ancient method of making steep-sided cultivation ridges, weeding growing crops became more difficult, and new implements for harvesting grain, such as scythes or reaping machines, could not be used.

The speed with which Irish farmers adopted some new techniques showed that they were not irrationally conservative, but welcomed changes which had proved to be real improvements. In the early decades of the century, new plough types almost entirely replaced locally designed implements, and in the middle decades, reaping machines and threshing and winnowing machines were enthusiastically introduced into many parts of Ireland. The widespread use of very large machinery, such as portable steam powered threshing machines, shows that the scale of investment required to buy such large equipment did not necessarily limit their use. In this case, farmers hired the machine for a few days each year. In other instances, the implement was often borrowed from a better-off farmer, in return for an agreed amount of labour.

Evidence that Irish farmers were much better cultivators than contemporary experts recognised, can be seen in the crop yields they produced, on both individual patches, and in the country as a whole. One debate which was carried on in farming literature, centred on whether the 'improved' method of growing potatoes in drills was better than the 'common' method of growing them in ridges. Improvers listed all the advantages of drill husbandry, apparently demonstrating the undeniable superiority of the method. The debate frequently ended, however, with an admission that ridge cultivation produced higher yields!

In Ireland as a whole, output from arable farming increased during the first half of the nineteenth century, and particularly during the Napoleonic Wars (1803-1815). By the 1820s, Ireland, for all the alleged inefficiency of its farmers, was being called 'the granary of Britain.' The expansion in cultivation continued until the middle of the century. Rich farmers in areas such as the Golden Vale of Tipperary, north east Leinster, or the newly reclaimed land along Lough Foyle, continued to produce large quantities of grain and vegetables. In terms of area under crops, however, the biggest increases took place in parts of the country which had previously been described as 'waste' or 'marginal' hill and bogland. Most of this expansion was the result of the labour of the rural poor, whose numbers were increasing rapidly. These people 'reclaimed' moorland and bog for crops, using hard labour and well-tried skills. The crop which allowed this spread was the potato. It was estimated that a family of eight could live for a year on the potatoes produced from one Irish acre of ground.

By 1845, the landscape of Ireland was densely populated and intensively cultivated. This was particularly true in areas of marginal land. In these areas, however, poor people were increasingly living on a knife-edge. Landlords or farmers would allow families to settle on hilly land and reclaim it, for a period during which rent was either not charged, or charged at a reduced rate. When the land had been upgraded by digging, liming, manuring and cropping with potatoes and oats, the rent would often be increased. In the most notorious cases, it was alleged that landlords would move the families higher up into the hills to begin the reclamation process all over again, but on even poorer land. The arrival of the potato blight (*Phytophthora infestans*) led to the destruction of the staple food of these impoverished masses, and decimated the population. In the wake of the suffering, crop cultivation receded back down the hillsides, leaving only the traces of old cultivation ridges, which can still be seen on most Irish hillsides today, a memorial to courage, hard work, and eventual destruction. The overall decline in tillage between 1855 and 1901 can be seen clearly in the following figures of crop acreages.

YEAR	OATS	MEADOW	WHEAT	BARLEY	FLAX	POTATOES
1855	2,118,858	1,314,807	455,775	226,629	97,075	982,301
1901	1,099,335	2,178,592	42,924	161,534	55,442	635,321

(Ireland, Industrial and Agricultural, P.308)

CASSIE AND PADDY McCAUGHEY MAKING RIDGES IN
CO. TYRONE IN 1906
Spadework often involved making ridges built up by turning sods over
onto untilled strips. These ridges were known as lazy beds or *iomairí*.
They were an efficient way to provide surface drainage in heavy soil,
and also provided deeper soil for plants in areas where top soil was
thin. During the nineteenth century, rich farmers found that employing
teams of spadesmen was a very effective way to break in previously
marginal land for cultivation. (Rose Shaw Coll.)

THE McALLISTER SISTERS SETTING POTATOES ON RIDGES, GLENSHESK, CO. ANTRIM, 1920s

Techniques such as spadework were described as 'common' during the nineteenth century. This usually implied that they were inefficient. In fact, while they were slow and labour intensive, techniques such as spadework or reaping with sickles were very refined. (Welch Coll.)

SPADESMEN IN WATERFORD *c.* 1824

There were more types of spade in Ireland than any other farm implement. In 1830, one spade mill in Co. Tyrone was producing 230 different types. Regional patterns can be seen in spade design. In the west of Ireland, one sided spades known as loys (*láighe*) were used. Elsewhere two-sided spades were common, but their blades tended to become longer and narrower further west.

PLOUGHING BY THE TAIL

This caricature, published in 1805, is an inaccurate representation of the most notorious Irish farming practice, where horses were attached to implements by tying them to their tails. The practice was made illegal in 1631, but was reported frequently in the eighteenth and early nineteenth centuries, and was alleged to have continued in isolated cases, as late as the middle of the present century.

Ploughing by the tail was no more cruel than many practices praised by improvers, but it was certainly inefficient. Even poor Irish farmers had an alternative in straw harness (*súgán*) which could be made at home from twisted or plaited straw, and was recognised as efficient and humane, as well as cheap.

COMMON IRISH PLOUGH IN USE NEAR HILLSBOROUGH, CO. DOWN IN 1783

This illustration by William Hincks shows one of the heavy wooden ploughs which continued in use in many areas until well into the nineteenth century. They required four, or sometimes six, horses for draught, and up to three men to operate them. The ploughman guiding the plough was assisted by another man who led the horses from the front. A third person, often a young boy, was sometimes required to lie on the beam to keep the plough in the ground. The ploughs were criticised by improvers because so many horses and men were required to operate them, and also because the furrows they turned did not lie flat on the ground. This was probably an advantage on undrained ground, however, where flat furrow slices would have aggravated waterlogging.

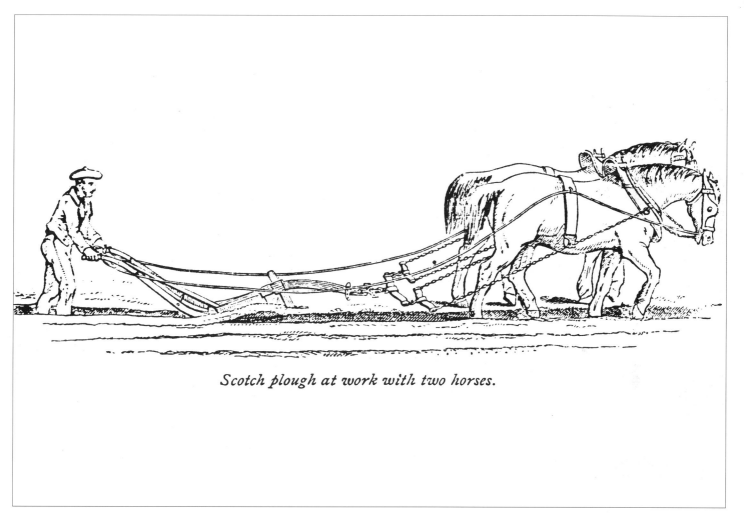

Scotch plough at work with two horses.

SCOTTISH SWING PLOUGH, AROUND 1871

This illustration shows the metal plough most approved for Irish conditions by improvers. Irish farmers were quick to adopt the new designs, and by 1830 the older wooden ploughs were becoming rare. These ploughs, which could be operated by one man and two horses, required great skill from the ploughman, who had to hold the plough at a constant depth and turn a furrow of constant width. However, they could be made to respond very quickly to bumps in the ground or changes in slope, a great advantage in hilly areas.

PLOUGHING WITH OXEN, BALLYGIBBON, CO. CORK, *c.* 1916

Since the later Middle Ages, most Irish farmers have relied on horses for draught. However, throughout the nineteenth century, there was an ongoing debate in agricultural journals, as to whether oxen might not be better. The debate was condemned as 'theoretic' by 'practical' farmers, but some gentry used oxen on their estate farms. This rare photograph taken on Sir Eustace Beacher's estate, shows Kerry cattle which were used for draught purposes, at least since the 1770s.

STEAM PLOUGH, c. 1860

Complex technology does not necessarily mean improved efficiency. Steam ploughing, which appeared on very large farms in the mid-nineteenth century, was mechanically ingenious, but by 1875 it was estimated that there were only seven steam ploughing complexes in the whole of Ireland, as opposed to 389 steam powered threshing mills.

At first, writers were very enthusiastic about the new technology, but by the early twentieth century, recognised that it was impractical. In this case, the caution of Irish farmers in adopting all new techniques was shown to be well justified.

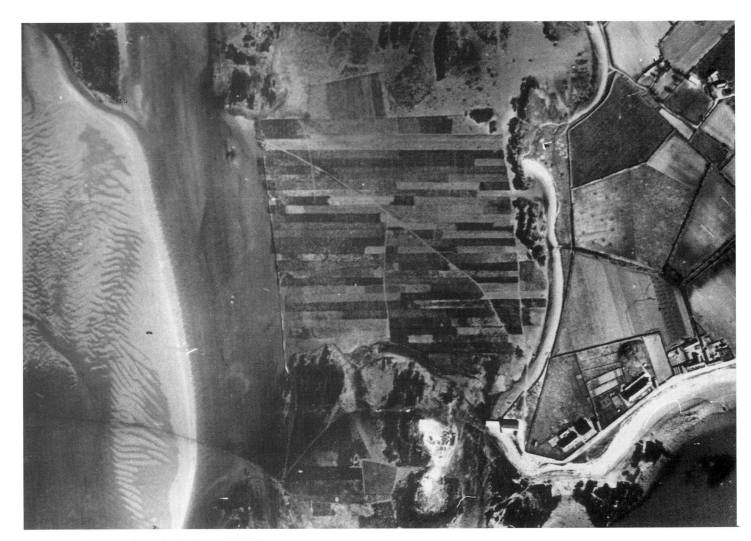

SEAWEED BEDS IN CARLINGFORD LOUGH

The use of seaweed as a fertiliser in Ireland has a long history, and by the later eighteenth century, coastal farms were let at higher rents because of access to this source of manure. In sandy areas, stones were often laid on the sea-bed to encourage weed to grow. On highly organised estates, these stones looked like underwater field boundaries, and rights to cut weed within particular 'fields' were included in leases.

CUTTING SEAWEED, INISHERE ISLAND, CO. GALWAY, *c.* 1930
In rocky areas, or where weed was plentiful, demarcation of cutting rights was much less defined. After a storm, when weed was washed ashore, everyone able for the work would gather to collect it. In spring the weed was often cut at low tide, as in the photograph shown here.

Men also went out to collect the weed in boats, often floating it in as one large heap or raft behind the boat. Sea weed was often spread over potato ground where its effects lasted for about one year. (Mason Coll.)

CARRYING ANIMAL DUNG IN A CREEL, IN 1900
Improvers claimed that Irish farmers did not make their manure heaps properly, as liquid manure was often allowed to seep away. However, small farmers knew that an adequate supply of animal manure could make the difference between survival and starvation. Without the manure of at least one cow, potato yields would not be big enough to feed a family. Manure heaps were often placed outside cottage doors, or in the case of byre dwellings, the heap was often allowed to build up in the house itself. One account from Gweedore, Co. Donegal, in 1838, claimed that up to fifteen tons of dung might be removed from a byre dwelling in spring!

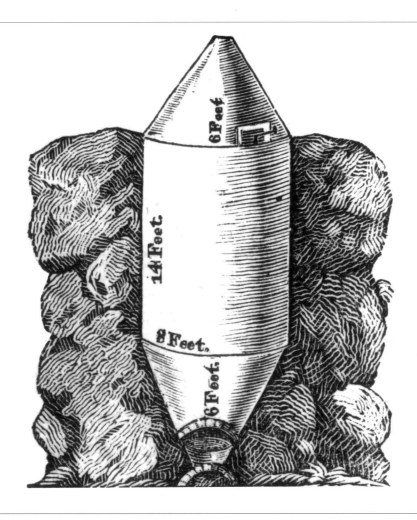

SUGGESTED PLAN FOR A LIME KILN, CO. KILDARE 1807

A lot of Ireland has limestone as bed rock, and by the late eighteenth century the value of burnt limestone, marl, and even seashells as fertiliser was well understood. In fact, by the nineteenth century, over-liming of ground was being identified as making land unsuitable for growing oats, for up to forty years. This was an unusual instance where Irish farmers were over-enthusiastic in their use of a practice advocated by improvers.

HARROWING ROCKY LAND

Harrows are used to break up clods on newly-tilled ground, or to mix seed and soil together. Improvers criticised 'common' Irish harrows, claiming that poor construction meant that the spikes or tines on the underside of the harrow frame often followed one another in lines rather than each breaking up a separate piece of ground.

However, this excellent drawing shows that the harrow was attached to the horse off-centre, which meant that tines would not follow directly behind one another. The drawing also shows *súgán* harness, made from plaited straw, a 'common' technique which was both cheap and highly efficient.

SOWING SEED BROADCAST

One of the earliest projects undertaken by the Dublin Society was the publication of Jethro Tull's *Horse Hoeing Husbandry* in 1733. His key invention, the seed drill, provided a mechanical way of sowing grain seed evenly. Throughout the nineteenth century, however, most small Irish farmers continued to sow grain by hand. It is very easy to throw seed on the ground. It is very difficult to sow it so that it spreads out evenly over the surface. People skilled at this were much in demand when crops were being sown in spring.
(Green Coll. WAG 262)

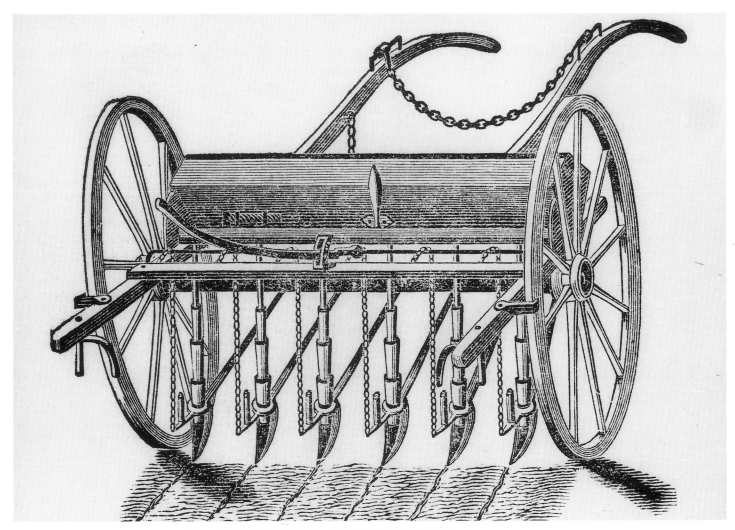

HORSE DRAWN SEED DRILL MADE BY RITCHIE'S OF ARDEE, CO. LOUTH IN 1847

Some Irish farmers cultivating a large acreage of grain crops did find it worthwhile to invest in a horse drawn seed drill. Seed fell from the box, down the pipes and was sown in long, straight lines, evenly spread over the field. Irish foundries made seed drills, but late in the century, light American implements were becoming popular.

MAKING RAISED DRILLS FOR POTATOES, IN 1871

This illustration, from a classic Scottish farming text, shows how drill cultivation allowed systematic preparation and planting of potatoes. The ploughman on the right, using a two sided drill plough, opens up long straight rows in earth which has already been well broken up. Teams of women follow, spreading manure and setting the potato seed. The ploughman on the left follows the women, closing the heaped up earth over the seed.

Drill cultivation was hailed as one of the triumphs of the 'scientific' approach to farming. The long, straight, equidistant rows of plants meant that horse-drawn equipment could be used for weeding between them, and eventually, that harvesting could also be mechanised. The system quickly became popular in many wealthier areas of Ireland, where, by the 1830s, it was claimed it had almost replaced the cultivation of potatoes in ridges, seen at the beginning of this section.

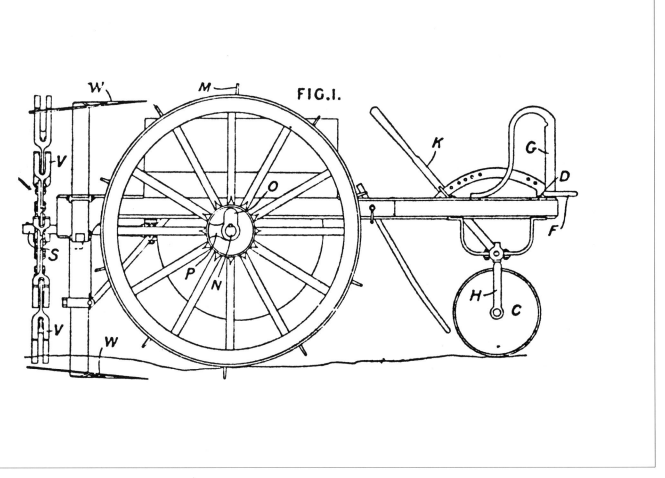

HANSON'S POTATO DIGGER

This invention, patented in 1852 by James Hanson of Doagh, Co. Antrim, was probably the most important Irish contribution to 'improved' farm technology during the nineteenth century. The machine became a prototype for diggers manufactured in many countries. It could only be used for potatoes grown in raised drills.

When the digger was pulled by horses along a raised drill, a large blade cut underneath the tubers growing in the drill. The iron forks at the back of the machine revolved as it was pulled, and these kicked the potatoes out of the drill. They were left lying on the surface of the ground where they could be collected easily.

ADVERTISEMENT FOR POTATO SPRAY

Potato blight is a fungus which attacks the leaves of the plant. From there it can be washed onto tubers by rain. A spray for killing the fungus was developed in France in 1882. The Bordeaux fungicidal mix, and later the Burgundy mix, were made up of copper sulphate and washing soda. By the 1890s, there was a major drive to introduce spraying throughout Ireland. This early twentieth century advertisement by Harringtons of Cork, stresses the benefits of spraying potatoes.

A MACHINE FOR LIFTING HAY, CO. KILDARE, 1807

In 1800, hay was not widely grown in Ireland. The mildness of the climate meant that housing and feeding of livestock during winter was not common. Improvers criticised Irish methods of hay making, claiming that Irish farmers cut grass too late, and left hay in the fields too long. Both practices, they claimed, meant that nutrients were washed out and that up to twenty per cent of hay rotted in the field. Irish farmers, however, pointed out that grass in Ireland was more luxuriant than in England, and that it was much more likely to overheat in stacks. In extreme cases, overheating could cause spontaneous combustion.

A FIELD OF HAY LAPS IN WESTERN IRELAND, c. 1900
Despite the relatively recent history of hay making on small Irish
farms, ingenious local methods were developed to save hay from
rotting in the damp climate. One of the most widely praised of these
was the making of hay laps (*gráinneog*), small rolls of hay described by
Victorian agriculturalists as being about the same size as a lady's muff.

The round shape of the laps meant that rain tended to run off, rather
than sink into the roll, while the hole through the centre meant that the
hay dried quickly. It was claimed that hay could survive in laps for up
to eight days of incessant rainfall.

HAYMAKING ON THE CLONBROCK DEMESNE, AHASCRAGH, CO. GALWAY, *c.* 1900

Large scale haymaking had become highly mechanised by the end of the nineteenth century. The horsedrawn hay rakes shown here had teeth, or tines, which could be raised or lowered using a lever beside the driver's seat. When the rake had gathered up a large amount of hay, the teeth were raised, leaving the hay in a heap. (Clonbrock Coll.)

CARTING HAY

By 1900, hay was the most widely cultivated crop in Ireland. The increase can be related to the shift from arable to pastoral farming, and to more intensive farming which led to an increase in the practice of housing livestock during winter. (Green Coll. WAG 1956)

PULLING FLAX NEAR CUSHENDUN, CO. ANTRIM *c.*1915
Flax cultivation spread with the development of the linen industry during the eighteenth century. This industry was increasingly concentrated in Ulster. During the nineteenth century, however, local cultivation of flax declined even in the north, and flax used in the linen industry was mostly imported. Flax required well prepared ground, and pulling it by hand was also labour intensive. Some farmers were reluctant to grow the crop because it was such hard work, and also because it took so many nutrients from the soil. Agriculturalists advised that flax should not be grown on the same patch of land twice within seven years. (Green Coll. WAG 1011)

REMOVING FLAX FROM THE DAM NEAR CUSHENDUN, *c.*1915

After pulling, flax was steeped in long, narrow dams filled with stagnant water. Sheaves or 'beets' of flax were usually placed in two layers under the water, and weighted down with stones. The length of time flax was left in the dam depended on the water temperature.

Within one or two weeks the woody central part of the flax stalk, known as 'shous,' had rotted, which made it easier to separate from the linen fibres, after the sheaves had been dried. This photograph was taken on the McSparran farm at Cloney. (Green Coll. WAG 1062)

'HANDSOME KATE KAVANAGH' FROM THE BARONY OF FORTH, CO. WEXFORD IN 1841

This illustration shows a toothed sickle (*corrán cíorach*), the most common tool used by Irish farmers for harvesting grain during the early decades of the nineteenth century. Both men and women used sickles, which cut grain slowly, but very neatly. The reapers held the crop in one hand while they cut through the stalks. This meant that seed was not shaken out of the heads and that straw could be cut very near the ground. A lot of weeds could also be left uncut, ensuring a cleaner crop.

PADDY STOREY REAPING OATS, NEAR TOOMEBRIDGE, CO ANTRIM *c.*1920

Irish reapers often worked in teams across a ridge, and each pair of reapers was usually followed by a binder who made bands of straw and used them to tie the grain in sheaves. Teams of Irish harvesters who travelled to England in the nineteenth century, actually slowed the introduction of scythes to the grain harvest in areas such as Yorkshire.

Irish labourers were paid very low wages and by hiring a large team, English farmers could have their crops harvested as quickly, and much more neatly, than those mown by local scythesmen.
(Green Coll.WAG 298)

SCYTHING OATS, 1890s

Scythes were probably introduced to Ireland by the Normans, who used them for haymaking. During the nineteenth century, they were more frequently used by small farmers, as hay became an important crop. They were not used widely in the grain harvest, however, until late in the century. A scythesman could cut grain up to four times faster than a reaper using a sickle, but the crop was shaken much more during cutting, leading to a loss of seed, and the straw was cut less neatly. (Lawrence Coll. C5716)

FLAILING OATS, TORY ISLAND, CO. DONEGAL, *c.* 1930
In Ireland, flails are made from two sticks tied loosely together. The person using the flail held one stick and used the other to beat a sheaf of grain laid on a board on the ground. The method was used throughout the nineteenth century, even after the introduction of mechanical methods of threshing. It was seen as a way of providing winter employment for servants and labourers who might otherwise be idle. (Mason Coll.)

LASHING OATS, ARAN ISLANDS, CO. GALWAY *c.* 1930
An ancient method of removing grain seed from the head of a sheaf was simply to bash it against a stone or a piece of wood. During the nineteenth century, however, the technique was increasingly used if only small quantities of grain were required, either to feed poultry or to make a loaf of bread. (Mason Coll.)

PUBLIC TRIAL OF McCORMICK'S REAPER, VIRGINIA, USA, 1831

Reaping machines cut grain mechanically. A successful reaper had been developed by a Scotsman, Patrick Bell, in the late 1820s, but it was not until the Americans McCormick and Hussey brought their machines to Europe in the early 1850s that the importance of their invention was recognised. American machines were shown in Belfast in 1852, and Dublin in 1853.

TWO HORSE REAPING MACHINE CUTTING OATS, NEAR TOOMEBRIDGE, CO. ANTRIM, *c.* 1920

By 1895 it was estimated that there were almost 15,000 horse operated reaping machines in Ireland. Many farmers who could not afford to buy a machine for themselves often obtained one from a wealthier neighbour, either by paying cash or in exchange for labour. Irish foundries, such as Pierce of Wexford, manufactured reapers of different sizes which could be operated by one horse or two. (Green Coll. WAG 1158)

SHERIDAN'S THREE-HORSE POWER THRASHING MILL.

HORSE OPERATED THRESHING MACHINE MADE BY SHERIDANS OF DUBLIN, 1848

Machines for threshing grain were developed in Scotland in the eighteenth century. By 1800 a few wealthy farmers were installing water powered machines on farms in Ireland. It was not until the 1840s, however, that Irish foundries began to manufacture efficient horse operated machines. A revolving drum inside the machine beat the grain out of sheaves as they were fed through. The machine shown here would have been installed inside a barn, but operated by horses walking round a gearing wheel outside. By 1875 it was estimated that there were almost 10,000 horse powered threshing machines in Ireland.

PORTABLE THRESHING MACHINE

Steam powered threshing machines were introduced to Ireland in the early 1850s. By the end of the century they had become a common sight in many parts of the country. Farmers hired the machine for several days and with the help of neighbours, threshed all the grain on the farm. This new way of using neighbours' help was a case when mechanisation made farmers more dependent on one another. Many examples of mechanisation were thought to make farmers more independent and therefore to hasten the break-up of communities.

WINNOWING IN THE OPEN AIR, *c.* 1930s

Winnowing removes the shells or husks from grain. The method shown here involved pouring the grain seed from a tray out of doors on a breezy day. The heavy kernels fell onto a sheet spread on the ground, while the light shells blew away. Some areas had 'shelling' hills, where breezes were common and people would gather to winnow their grain. The trays used were made of skin stretched around a wooden frame. In southern parts of Ireland the tray was known as a bodhran and was similar in construction to the one-sided drum of the same name which is used in traditional music. In the north, the trays were also known as wights or wechts. (Mason Coll.)

WINNOWING MACHINE IN USE, c. 1871

Machines for winnowing grain were developed in Holland and Scotland during the eighteenth century. They were quickly adopted in Ireland during the nineteenth century. By 1875 there were more than 2,000 in use. Most were manually operated. The handle on the side of the machine turned wooden fans which set up a draught of air. Seed was poured into the top of the machine and fell through a series of vibrating wire mesh riddles. The draught blew the husks out of one end of the machine, and the mesh riddles cleaned the seed.

APPLE ORCHARD, CO. ARMAGH, *c.* 1905
Some crops were very closely associated with particular counties.
Apples were widely grown in Armagh and Tipperary, and cider was
made on a small scale in both counties.

WOMEN SELLING VEGETABLES
Farm gardens provided families with vegetables such as onions, carrots
and cabbage. Excess produce was sold, often at a local market or fair.
(Lawrence Coll. R2703)

CATTLE FAIR, DUNGARVAN, CO. WATERFORD

Even before the shift from arable to pastoral farming, livestock were kept on almost all Irish farms. On small farms, cows were particularly important, as sources of milk, meat and dung for fertilisers. Between 1850 and 1900 cattle increased in numbers from just under 2.5 million to more than 4.5 million. The number of cattle exported to Great Britain had increased from just over 240,000 to more than 642,638. (Lawrence Coll. R3583)

LIVESTOCK

If developments in Irish farming during the first half of the nineteenth century were centred on crop production, the second fifty years was a period during which livestock husbandry was transformed. A long-term swing away from arable farming began around the time of the Famine, and numbers of almost all kinds of livestock increased. The following figures show numbers of livestock kept per 1000 acres in Ireland in 1855 and 1900:

	CATTLE	SHEEP	PIGS
1851	143	102	52
1901	230	215	60

(*Ireland, Industrial and Agricultural*, p320)

Irish methods of farming livestock were criticised as much as techniques of crop production. Some local breeds of livestock, such as Kerry cattle and Roscommon sheep were seen as useful, but most native types of farm animals were mentioned dismissively, if at all. Experts urged that new breeds of livestock, particularly those developed in England and Scotland, should be introduced as quickly as possible. To their frustration, Irish farmers were highly selective in accepting the new livestock. However, when we look at farming conditions in Ireland this caution is easily understood. As with crop production, local methods of livestock husbandry made more sense than outsiders recognised. The most successful local breeds were 'dual-purpose', which made them well suited to the small scale of most Irish farms. English improvers praised huge work horses such as Shires or Clydesdales, but Irish Draught horses, which were much smaller, standing just over fifteen hands high, were more efficient on small or medium farms. Not only could they be used for tillage, they were also excellent for driving and riding. Kerry cattle were not so physically impressive as the big specialised breeds developed in Victorian England, but they were hardy and produced both good quality milk and beef. They were rightly praised as 'the poor man's cow.'

Agricultural writers were horrified at how closely people and animals lived together on many small farms in Ireland. On very poor farms, people and animals often ate and slept in the same living space. Cattle and pigs were most commonly kept indoors, and 'the pig in the parlour' became a music-hall image of the typical Irish farm. The practice was a rational one, however, as pigs flourish in the same sort of temperatures as human beings. The same seems to have been believed about cattle, particularly cows which were near calving. The untimely death of a large animal could leave a poor family destitute, and an obsessive concern with the welfare of livestock is understandable. Throughout Ireland, better-off farmers often had well-planned farmyards, sometimes laid out according to plans declared ideal by improvers. By 1900, even small farms usually had purpose-built outbuildings for livestock. Stall feeding of animals housed during winter became more systematic, with the increase in the cultivation of hay and other fodder crops.

Attempts to introduce new breeds of livestock became more successful during the later nineteenth century. Local farming societies gave prizes or 'premiums' for high quality animals displayed at annual shows. Some landlords bought pedigree male animals such as bulls, rams and stallions, and tenants were sometimes allowed to cross their own livestock with these. Towards the end of the century, bodies such as the Congested Districts Board began to distribute pedigree breeding stock, including cockerels, to small farmers. By 1900, Shorthorns had become the typical cattle on Irish farms, and Scottish Blackface and Border Leicester sheep had become common. Some Irish breeds survived, and even flourished, however. Kerry cattle remained popular and a register of Irish Draught stallions was begun in 1905. Large White Ulster pigs were found to be suitable for slaughtering on the farm and dry curing as practised in the north of the country, and a herd book for the breed was established in 1907.

The increase in livestock numbers was particularly dramatic for poultry. By 1900s number of poultry had multiplied to over 18.5 million, more than three times their number in 1850. This last increase was particularly important for farm women, who looked after the poultry, and could sell or barter the eggs produced. The shift to livestock farming, and the accompanying sharp decline in tillage meant that by 1900 the Irish farming landscape had changed dramatically. The 'granary' had become a leading pastoral farming region.

POLLED IRISH COW, FROM A PAINTING BY WILLIAM SHIELS, *c.* 1848

Hornless (*maol*) cattle have a very long history in Ireland, and cattle of the type shown here were still said to be common in the mid-nineteenth century. By 1907, however, it was claimed that they were extinct. They were reported to be as large as Shorthorn cattle, and a good dairy breed. In 1831, the agricultural writer, J. C. Loudon claimed that dairying was the best managed part of Irish husbandry. He identified the main dairying counties as Kerry, Cork, Waterford, Carlow, Meath, Westmeath, Longford, Fermanagh and the mountains of Leitrim and Sligo. The best butter, he claimed, was made in Carlow, and the worst in Limerick and Meath.

IRISH LONGHORN COW, IN 1834

This large breed was found mostly in Tipperary, Limerick, Meath and Roscommon. In 1834, an English writer described them as 'most valuable' animals. The origins of the breed are not clear, but horns and skulls showing similar characteristics have been found on early Christian sites. During the nineteenth century, some large landowners crossed the breed with English Longhorns, and by the end of the century Irish Longhorns were no longer recognised as distinct.

A KERRY HEIFER 'MONINA', EXHIBITED AT THE R.D.S. SPRING SHOW, DUBLIN, IN 1887

Agricultural shows encouraged farmers to improve their livestock. Kerry cattle had achieved a high reputation for hardiness and productivity by the late eighteenth century. They were the most famous of the dual purpose Irish breeds of cattle kept on small farms both for milk and beef. A register of Kerry cattle was established in 1887 by the Royal Agricultural Society of Ireland, and a herd-book was established in 1890.

COW AND BULL
The small cow shown here is of the Dexter breed. Dexter cattle
probably have the same origins as the Kerry breed, and at the end of the
nineteenth century the problem of whether to establish a separate
herd-book for Dexters was being debated. This photograph is a witty
comment on the effects of selective breeding in cattle, producing very
small, as well as very large animals.

MRS McCOLLUM MILKING A SHORTHORN COW, CUSHENDUN, CO. ANTRIM (Green Coll. WAG 1952)

SHORTHORN HEIFER AT R.D.S. SPRING SHOW, 1915

By 1900, Shorthorns were the most common breed in Ireland, and throughout large areas of north-west Europe. They were a dual-purpose, improved breed, and so fitted well into the economy of small Irish farms. This photograph shows Jack Walker of Co. Wexford, with his father, William Cunningham Walker.

BUTTERMAKING, CO. TYRONE, c.1906

Butter was made on most small farms; in the north often from the whole milk. Women often had their own techniques for making butter, but most surviving accounts emphasise the need for cleanliness, the careful souring of milk in correct, moderate temperatures, and the necessity of washing buttermilk out of freshly made butter. It is unlikely that churning often took place out of doors, as shown in this photograph. Difficulties in lighting a cottage interior may explain why the photographer, Rose Shaw, had her subject pose outside. (Rose Shaw Coll.)

ROADSIDE BUTTER MARKET, CO. CORK, 1890s

The unorganised system of trading farm produce led to problems. Leaders of the co-operative movement claimed that this was especially true in the case of the butter trade. Women were sometimes tempted to hold on to butter until prices rose, and the shopkeepers and butter merchants who bought the butter also sometimes kept it for the same reason. This meant that butter which was good quality when made, was often inedible on arrival at urban centres. In cities such as Cork and Liverpool, a lot of Irish butter was used only to grease machinery!

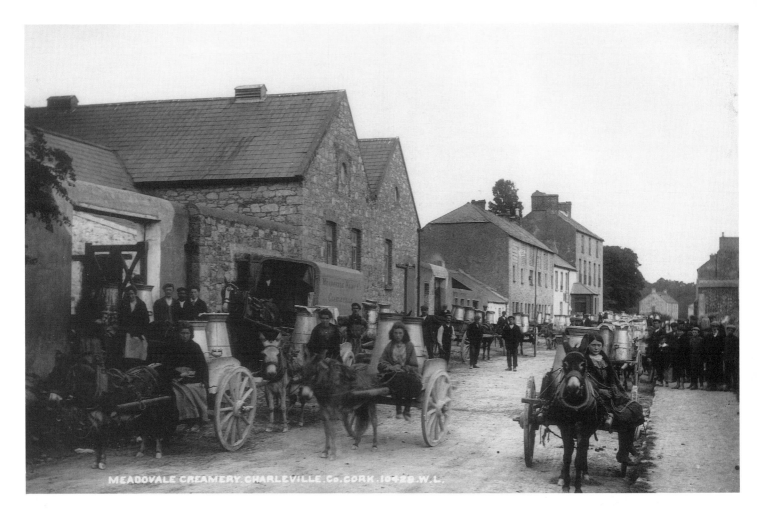

MEADOWVALE CREAMERY, CHARLEVILLE, Co. CORK

MEADOWVALE CREAMERY, CHARLEVILLE, CO. CORK

By the 1890s the reorganisation of the Irish dairy industry had begun. A major impetus came from the co-operative movement (Irish Agricultural Organisation Society) which set up creameries that bought milk from farmers, processed butter to a high standard, and sold it in large urban centres. Irish butter achieved its present high reputation because of this movement, but farm women lost out. Men brought the milk to the co-op and received the 'creamery cheque'. Women, therefore, lost control of a significant part of the farm income. (Lawrence Coll. R10,428)

BUTTER MARKET. CORK. 6976. W.L.

CORK BUTTER MARKET

Not all trading of Irish agricultural produce was small-scale. Urban centres such as Dublin, Belfast and Cork had large markets well before the nineteenth century. Cork Butter Exchange was founded in 1730.

The large scale inspection and classification of butter at the exchange continued until 1925. (Lawrence Coll. C6976)

SHEEP FAIR, ATHLONE, CO. WESTMEATH
Between 1850 and 1900, sheep numbers had increased from just under
2 million to almost 4.5 million, and sheep had spread over much of the
marginal hilly land which had been cultivated before the famine.
(Lawrence Coll. R3583)

WICKLOW SHEEP

In 1855, these sheep were described as 'wild little animals, without horns, and with white faces and legs... They are larger towards the base of the mountains, where the pasturage is better, and the wool is tolerably fine and rather long, though mixed with hair.' Wicklows lambed early in the year, and the ewes were said to be excellent mothers. However, by the 1850s they were being crossed with larger breeds, particularly Cheviots, and Wicklows were disappearing as a distinctive type. In 1902, Wicklow-Cheviots were claimed to be the best mountain sheep in Ireland.

KERRY HILL SHEEP

These sheep were larger than the Wicklow type, but in 1855 were described as 'wild, coarse, unthrifty animals'. They were black, brown and white in colour, and had fleeces which were hairy on the back, but short and fine at the sides. It was claimed that Kerry sheep were slow to mature, and difficult to fatten, but that their mutton was of good quality. They were great favourites with butchers, because they had a large proportion of loose fat. In the mid-nineteenth century, however, one agriculturalist concluded that 'the sooner some more profitable breed is substituted for the old, the better for all parties.'
(Paintings by William Shiels)

ROSCOMMON SHEEP

In the early nineteenth century, several distinctive types of sheep were identified in counties Galway and Roscommon, and Roscommon sheep were claimed to be descended from these. In 1877, the breed was described as 'recently established', having arisen from crossing native western sheep with the Leicester breed. Roscommons were described as 'big useful sheep, though lacking in symmetry'. They were said to be slower to mature than other long-woolled breeds, but to have a white, bright and lustrous fleece.

SPINNING WOOL, INISHOWEN, CO. DONEGAL

The home production of wool, knitting and weaving, were established in Ireland in ancient times. By the end of the nineteenth century it was estimated that more than one sixth of a small farm's income might come from money earned by women through knitting and sewing. By this time, however, local merchants usually provided the women with wool which had already been spun. (Green Coll. WAG 1181)

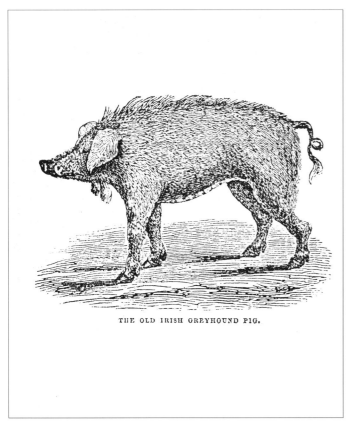

THE OLD IRISH GREYHOUND PIG.

THE IMPROVED IRISH PIG.

GREYHOUND PIG

In 1847, 'old Irish' pigs were described as 'tall, long-legged, bony, heavy-eared, coarse-haired animals, their throats furnished with pendulous wattles, called in Irish *sluiddeen*, and by no means possessing half so much the appearance of domestic swine as they do of the wild boar.' They were said to be remarkably active and could 'clear a five barred gate as well as any hunter'. By 1847, they were to be found mostly in the west, and especially in Galway. The last reported sighting of Greyhound pigs came from Liverpool, where it was claimed that, during the Second World War, an Irishman had kept a herd living off a city rubbish tip!

IMPROVED IRISH PIG

Fat pigs had become common in Ireland by the mid-nineteenth century. It was claimed that selective breeding, and crossing with pigs of the Hampshire and Berkshire breeds had produced the change. The pig in this illustration won a prize at the R. D. S. Show in 1847. It weighed forty-one stones and belonged to a 'humble cottier', Peter Flood. Like other pigs of the same kind it was said to fatten easily, while keeping the rich flavour of the Old Irish pig. (*Irish Farmers Gazette*, 1847,p57.)

CASTLEWELLAN FAIR, CO. DOWN

The Large White Ulster pig was one of the most successful Irish breeds. This photograph shows a Large White Ulster pig for sale. It was fairly unusual to sell fattened pigs alive, especially in Ulster. More commonly the pigs were slaughtered on the farm, and sold as dead meat. (Green Coll. WAG 1175)

INTERIOR OF A HOUSE, WITH A PIG, 1880
On small farms the pig was often known as 'the gentleman who pays the rent', as a pig fattened at the right time could provide the money required. However, pigs as well as people depended on potatoes, and pig numbers dropped dramatically after the Famine, from almost one and a half million in 1841, to just over half a million in 1848. By 1900 numbers had risen again, to over one and a quarter million.

A PIG KILLING, CO. ARMAGH

Pigs were killed for home use all over Ireland, and in Ulster, even pigs reared for sale were usually killed on the farm. The method generally involved stunning the pig with a mallet, and then cutting its throat. The carcass was thrown into boiling water, and hair was scraped off. Afterwards, it was usual to hang the pig up and remove the innards; in this photograph, they are hanging on the wall. The heart and liver were eaten, and the intestines sometimes used to make sausages. The stomach, which is shown blown up at the feet of the boy on the left, was sometimes used as a football, but in some areas it was stuffed with potatoes, onions and oat meal. It was then boiled, sliced up, and fried. (Bigger Coll.)

77

HORSE FAIR, DUNDALK

Large horse fairs were well established by 1800. Numbers of horses did not increase greatly during the century. By 1901, there were about 564,916 horses in Ireland, an increase of just over 40,000 in the number recorded in 1855. Of these, 354,750 were used in farming. The decline in tillage during the second half of the century meant that there was no significant increase in demand for working farm horses. (Welch Coll.)

IRISH DRAUGHT HORSE, 1905

During the later nineteenth century, it became accepted that there were indigenous Irish work horses which differed markedly from English and Scottish improved breeds. In 1888, the Royal Dublin Society allocated an annual sum of £3,000 for the improvement of horse breeding in Ireland. Thoroughbred stallions were distributed to various parts of the country, and these served selected 'half-bred' mares belonging to tenant farmers. An annual register of thoroughbred sires was begun in the 1890s. After 1900, this work was undertaken by the newly formed Department of Agriculture and Technical Instruction. Stallions of 'the Irish Draught type' were listed after 1905.

CONNEMARA PONY, 1840, FROM A PAINTING BY WILLIAM SHIELS

Many parts of Ireland had ponies which were believed to have distinctive local characteristics. Some of these were probably direct descendants of native Irish horses (*gearráin*), and others the result of breeding between indigenous horses and animals brought in from elsewhere. Connemara ponies attracted the attention of improvers in the mid-nineteenth century because of their quality. Some writers claimed that there were strong similarities between Connemara and Spanish ponies, and that some of the latter must have been brought in to the area, possibly at the time of the Spanish Armada. (In folklore, so many features of western Irish life have been attributed to the influence of the shipwrecked Spanish Armada, that this short-lived disaster is credited with an impact as great as a mass colonisation!) During the last decade of the century, the Congested Districts Board did introduce stallions, of several breeds, in an attempt to improve the quality of local ponies. Whatever the outside influences, the modern Connemara has an international reputation as a distinctive, high-quality riding pony.

UNLOADING DONKEYS FROM A TRAIN AT WATERFORD, 1897
Donkeys only became common in Ireland during the eighteenth century, but by the mid-nineteenth century they had replaced working horses and ponies on many small farms, especially in the west. The number of donkeys in Ireland rose from just over 100,000 in 1850, to more than 240,000 in 1900. Some Irish donkeys were imported into England to provide seaside rides for holidaymakers. (Poole Coll. W. P. 822)

FEEDING POULTRY, THE MOY, CO. TYRONE, 1890
In 1900, it was estimated that the produce of twenty chickens equalled a cow in value. On even tiny farms, with only one cow, there might be a flock of one hundred chickens. The money made from their eggs was generally controlled by farm women. This meant that during the later nineteenth century, women on these small farms had a disposable income which rivalled that of their husbands.

DUCKS FOR SALE
Ducks were recommended to Irish cottagers because they required little care, ate insect pests, were good for eating, and produced large eggs. In the late nineteenth century, the breeds most recommended were the Aylesbury and the Rouen. By 1900 there were over 3 million ducks kept on Irish farms, almost 2 million geese, and well over 1 million turkeys. (Lawrence Coll. R6731)

EGG MERCHANT, CO. CORK, 1890s
Changes in the system of trading eggs in the late nineteenth century parallelled changes in the butter trade. For most of the century, women sold eggs to local shopkeepers, or egg merchants such as the one shown here. Women and traders were often tempted to hold on to the eggs until prices rose. This meant that eggs arriving at large urban centres were often inedible.

THE ANNUAL SHOW OF THE CLONBROCK AND CASTLEGAR CO-OPERATIVE POULTRY SOCIETY AT CLONBROCK, CO. GALWAY, 1902

This society was founded in 1898. Co-operative societies aimed to market clean, fresh, properly packed eggs, and to ensure that farmers obtained good prices for them. Unintentionally, however, the co-operative movement was partly responsible for removing the income obtained from eggs from farm women, to men. Many co-operative creameries developed egg marketing as a sideline. Men brought eggs as well as milk to the creamery. The money for the eggs was added to the milk cheque, which was given to the men. (Clonbrock Coll.)

A COMFORTABLY-OFF FARMING FAMILY, THE DANAGHERS FROM CO. LIMERICK, *c.* 1900

FARMING PEOPLE

During all of the great changes which occurred during the nineteenth century, most small and medium sized farms continued to be worked by the family who held the land. Men were the dominant decision makers, while farming wives had a subordinate role, assisting in all sorts of work, but also processing farm produce, for example, baking bread, curing pork, and making butter. Where the family had well-grown sons and daughters, there was only occasional need for outside assistance. However, at very busy times of year, in times of sickness, or where there were no children or very young children. it was common to ask for help from neighbours. Farmers often formed a partnership with a neighbour, exchanging labour on a day to day basis, or lending and borrowing horses or machinery. These exchanges built up long-term relationships which were known by many different names throughout Ireland including *comhar*, morrowing, neighbouring, working in means, joining and swapping. Groups of neighbours also helped one another, digging with spades, haymaking, reaping grain, or later in the century helping with the operation of a portable threshing machine. There were many terms used to describe these groups also, including, *meitheal*, boon, camp, *crinniú*, drag, gathering, and *sealbhán*.

Both wealthy and poorer farmers sometimes employed farm workers. It is very difficult to discover how many men worked as farm labourers, because even people with tiny holdings, often no bigger than a potato patch, described themselves as 'farmers' rather than 'workers'. Some farmers employed 'cottiers' who were given a house and the use of a patch of land paid for by labour. Despite difficulties in definition, however, it is clear that between the Famine and 1900, the number of full-time farm workers fell sharply. In 1841 it was estimated that there were 1,229,000 people employed as farm servants or labourers in Ireland. By 1891, this number had dropped dramatically, to 258,000. However, in many parts of Ireland, and particularly in the west, large numbers of farmers and younger members of their families continued to engage in migrant work, either or part of the year or for several consecutive years during their 'teens. These workers sold their labour in the richer parts of Ireland, and in many parts of England and Scotland. Spailpn (spalpeens), for example, commonly came from west and south-west Ireland and hired themselves throughout richer farming areas in Munster and Leinster for tasks such as spadework, reaping with sickles, or towards the end of the century as scythesmen. Large groups of workers from Mayo and Donegal travelled to Scotland as harvesters. Both men and women were employed, especially in the Scottish potato harvest, as 'tattie hokers.

In Ulster, hiring fairs were held twice a year in more than eighty towns. Young people, sometimes no older than seven years of age, were hired as farm servants by farmers for 'terms' of six months, receiving food and lodging and a small sum of money as payment. Servants were employed, not only on big farms, but also by poorer people, often elderly, who lived alone. The conditions servants experienced ranged, in their own words from 'slavery' to 'being treated like one of the family'. The hiring system was going strong by 1900, but elsewhere in Ireland, emigration meant that there were very few full-time farm labourers, farmers hiring casual workers only at harvest.

A FARMING FAMILY, CO. WEXFORD, 1905

This photograph shows the Corish family of Ring Sherane, Carne, south Wexford. They lived on the edge of Lady's Island and only a short distance from the sea. Their farm was about 10 acres. They collected seaweed from the shore, mixed it with sea sand and used it as a fertiliser. They had one horse and also did some work for local farmers. Their own small farm had a lot of outcropping granite boulders and was not very amenable to machine cultivation. Willows grew in nearby marshes near the edge of the lake. The father made sprays (spars) for thatching and also potato baskets as a sideline. By 1900 families like the Corishs were becoming typical on small farms throughout most of Ireland.

FOOTING TURF

Many tasks, such as harvesting turf, could require the participation of the whole family. On most farms, however, men were in control. They decided which crops and animals should be produced, and organised the rest of the family to carry out necessary tasks. Men usually undertook heavy tillage operations, using either spades or ploughs. The ability to work with horses was the most valued of all farming skills. Early in the nineteenth century, farmers often tried to divide their land between their children. This led to more, smaller farms with each succeeding generation. Landlords often resisted this practice, however, and after the Famine it had become very rare. A slow consolidation in farm size began. Most small mixed farms could be worked by two men, so later on in the century it became more common that one son would work closely with his father, learning essential skills and eventually taking over the running of the farm. (Lawrence Coll. R 2870)

BUTTERMAKING, CO. ANTRIM
Many women's skills were used to process farm produce.
Buttermaking, bread making, curing meat and cooking were all clear
examples of this.

HAYMAKING, 1890s

Some farming tasks were seen as particularly associated with women. The care of young livestock kept around the farm house, and milking cattle were two of these. However, on small mixed farms, any member of the family might be called on to help with any task. Women helped in the fields, especially at busy times of planting and harvest, and if there were no grown men in the family, could also be expected to work with horses.

A WORK GROUP IN A HAYFIELD, CO. TYRONE, *c.* 1906
Neighbours gathered in work groups at busy times of year, working in each other's fields in turn. Elderly or sick people, and clergy, were also often helped. A work group, could also be called to help not only in the fields, but to assist with major tasks, such as building a house.

Such a group might have only five workers, but could be very large indeed. In Ulster, just before the 1798 rising, it was claimed that 6,000 men met, insisting that they had gathered to dig a poor woman's potato patch. This was not believed! (Rose Shaw Coll.)

STAMPY PARTY, CO. LIMERICK, 1907-8
Farmers often had a special meal organised at the end of harvest, for
helping neighbours and hired workers. On this farm, stampy, made
from grated potatoes and seasoning, was made to celebrate the end of
the potato harvest.

CEILÍ

In rural Ireland, neighbours who worked together socialised together. It was very common for small farmers to ceilí together, often almost nightly. People would gather in a local house, to drink tea, tell stories, sing, dance or play cards, often until late at night. There was a high value placed on a person who was 'good crack,' that is who could perform well in conversation, whether gossip or stories.

Hiring Fair, Ballymoney.

HIRING FAIR, BALLYMONEY, CO. ANTRIM, *c*. 1910

The hiring of farm servants for terms of six months was well established by the eighteenth century. In the late nineteenth century, many small Ulster towns had fairs around the 12th May and 12th November. Young people looking for work would gather in the centre of the town, and farmers looking for servants would inspect them there. Each young servant would carry a bundle which contained clothes and other personal possessions. When a deal was made with a farmer on work, lodging and wages, the servant handed over the bundle and received a small amount of money, known as an 'earls' or 'earnest' in return. This made the contract legally binding. About four times as many boys as girls were hired as servants. Girls were usually given a place to sleep inside the farmhouse, but many boys slept in converted outhouses. Servants judged how they were treated mainly by the food they were given. A farmer who had a 'good meat' house, had no difficulty in finding workers.

A KERRY SPAILPÍN (SPALPEEN), *c.* 1868
Spailpín means 'a short spell', and these workers were generally hired for jobs which lasted only a few weeks. Most spalpeens came from Cork, Connemara and Kerry. Some went to England to work, but many also travelled within Ireland, particularly to counties Limerick, Tipperary, Waterford, east Galway, and parts of Leinster. In the later nineteenth century, spalpeens were most commonly employed as potato diggers.

TATTIE HOKERS IN SCOTLAND

Before the Famine, up to 60,000 migrant workers travelled from Ireland to Britain each year, often for harvest work. This painting by David Farquharson (1840-1907), shows workers in a potato field in Renfrewshire. The conditions endured by potato harvesters, or 'tattie hokers' were notoriously bad, especially in south-west Scotland.

Workers were housed communally in animal outhouses, and the lack of privacy, dirt, and absence of any space where wet clothes could be dried, horrified many contemporaries. The money obtained from migrant labour, however, was necessary for survival.

97

BLACKSMITH

It has been estimated that there were over one hundred trades and crafts practiced in nineteenth century rural Ireland. Some craftspeople, such as wheelwrights, saddlers, and blacksmiths were essential for farming.

Blacksmiths held a special position, as they were responsible for making or repairing most of the farm implements used before 1900.

SHOP, OMEATH, *c.* 1900

By 1900, shops were found not only in Irish small towns, but throughout the country. Shops provided a range of goods which could not be
manufactured locally and often engaged in barter as well as cash sales.
Many shopkeepers also provided credit, and farmers made use of this
to obtain fertilisers or seed at times when money was scarce.

The activities of some shopkeepers, however, earned them the name of
gombeen (*gáimbín*), as it was alleged that they used the system of
credit to exploit local people ruthlessly.

PUCK FAIR, KILLORGLIN, CO. KERRY

Fairs and markets were important centres of trade throughout the nineteenth century. Special fairs were sometimes arranged at spring or harvest, or on a date associated with a local saint. The Puck fair at Killorglin is still held between 10-12th August each year. A male goat is hoisted on to a raised platform at the beginning of the fair and remains there throughout. Socialising as well as dealing was intense at most fairs. (Lawrence Coll. R7640)

BIBLIOGRAPHY

Agricultural Classbooks [For Irish National Schools] 2 vols (Dublin, 1898)

Baldwin, Thomas **Introduction to Irish Farming** (London, 1874)

Bell J. and Watson M. **Farming in Ulster** (Belfast 1988)

Bell J. and Watson M. **Irish Farming - Implements and Techniques 1750 - 1900** (Edinburgh, 1986)

Bolder, Patrick **The Irish Co-operative Movement** (Dublin, 1977)

Doyle, Martin **A Cyclopaedia of Practical Husbandry** Rev. ed. W. Rham (London, 1844)

Evans, E. Estyn **Irish Folkways** (London, 1957 (1977))

Fitzpatrick, David 'The Disappearance of the Irish Agricultural Labourer, 1841 - 1912' in **Irish Economic and Social History** Vol 7 (1980)

Fussell, G. E. **The Farmer's tools,** 1500 - 1900 (London, 1952)

Gailey, Alan **Spade making in Ireland** (Holywood, 1982)

Hall, Mr and Mrs S. C. **Ireland, Its Scenery and Character** 3 Vols (London, 1841)

Hill, George **Facts from Gweedore** introd. E. E. Evans (Belfast 1971)

Ireland, Industrial and Agricultural (Dublin, 1902)

The Irish Farmer's and Gardener's Magazine Dublin (1832 - 1835)

The Irish Farmer's Gazette (Dublin, 1840 - 1849)

Library of Useful Knowledge **British Husbandry** 2 Vols (London 1834)

Loudon, J. C. **An Encyclopaedia of Agriculture** (London, 1831)

Low, D. **Elements of Practical Agriculture** (Edinburgh, 1854)

Micks, W. L. **An Account of The Congested Districts Board for Ireland** (Dublin, 1925)

Ó Danachair, Caoimhín 'The flail in Ireland' **Ethnologia Europaea Vol 4 (Antrim, 1971)**

O'Dowd, Anne **Meitheal** (Dublin, 1981)

O'Dowd, Anne **Spalpeens and tattie hokers** (Blackrock, 1991)

O'Muirithe, Diarmaid **A seat behind the coachman** (Dublin 1973)

Partridge, Michael **Farm tools through the ages** (Reading, 1973)

Purdon's Practical Farmer (Dublin, 1863)

Ryder, M. L. **Sheep and man** (London, 1983)

Salaman, R. N. **The history and social influence of the potato** (Cambridge, 1949 (1970))

Solar, Peter 'Agricultural productivity and economic development in Ireland and Scotland in the early nineteenth century', in **Ireland and Scotland, 1600 - 1850** eds T. M. Devine and D. Dickson (Edinburgh, 1983)

Sproule, John **A Treatise on Agriculture** (Dublin, 1839)

Vaughan, W. E. **Landlords and Tenants in Ireland 1848 - 1904** (Dundalk 1984 (1985))

Watson, M. 'Common Irish plough types and tillage practices' **Tools and Tillage** Vol 5.2 (Copenhagen, 1985)

ACKNOWLEDGEMENTS

Many people and institutions helped in the production of this book. They include Deirdre Brown, Fionnuala Carragher, Hugh Cheape, Leo Curran, Dermot Francis, Michael Maloney, Ann O'Dowd, Austin O'Sullivan, Maureen Paige, Vivienne Pollock, Ríonach Uí Ógáin, the photographic staff of the Ulster Folk and Transport Museum, the Ulster Museum, the Irish Agricultural Museum, the National Library of Ireland, the National Museum of Ireland, the Royal Dublin Society, The Department of Irish Folklore, University College Dublin, and Renfrewshire Museum and Art Galleries. Mervyn Watson was so closely involved with the book that it is impossible to assess the extent of his help. Without it, the book would not have been written.

SOURCES FOR ILLUSTRATIONS

Most line drawings in this book were taken from nineteenth century tours and agricultural texts. Some of the photographs date from the early twentieth century, but have been included because they show processes which were common before 1900.

PUBLISHED SOURCES

Anon. **A Right merrie and conceitede tour** (London, 1805), 25.

Barrow, John **A Tour round Ireland** (London, 1836), 14.

Berry, Henry F. **A History of the Royal Dublin Society** (London, 1915), 6.

Collins, M.E. **Ireland, 3 Union to the present day** (Dublin, 1990), 15, 34.

Curran, P.L. **Kerry and Dexter cattle** (Dublin, 1990), 62.

Fussell, G.E. **The farmer's tools** (London, 1952) illus. nos.29, 50.

Irish farmer's gazette (Dublin 1840 - 1849),13, 36, 52, 74.

Library of Useful Knowledge **Cattle** (London, 1834), 61.

Lovett, R. **Irish pictures** (London, 1880), 76.

Mason, Thomas H. **The islands of Ireland** 2nd ed. (London, 1931), 31, 49, 54.

The Illustrated London News, 2, 3, 4, 5, 93.

Pictures of Irish life (Cork, 1902), 32, 41, 67, 84, 90.

Stephen's book of the farm 2 vols. 3rd ed. (Edinburgh, 1871),27, 37, 55.

Thackeray, William Makepeace **An Irish sketchbook of 1842** in Works vol.18 (London, 1897), illus.7.

ARCHIVAL SOURCES

The following photographs and other illustrations are published by kind permission of the following institutions and individuals:

The Trustees of the Ulster Folk and Transport Museum, the Glass, Green and Rose Shaw Collections, 24,30,33,38,39,40,46,53,56,60,70,71,72,79,80,90,94,95.

The Trustees of the Ulster Museum, the Welch and Bigger Collections, 26,82,98,99.

The National Library of Ireland, the Lawrence, Clonbrock and Poole Collections.

The National Museum of Ireland, the Mason Collection.

The Irish Agricultural Museum, 65,86.

The Dundee Collection, 10,11.

Cambridge University, 20.

Renfrewshire Museum and Art gallery, 97.

The Earl of Roden, 1.

Mr R. Peters, 28.

Dr Caoimhín Ó Danachair, 92.